PUBLISHER COMMENTARY

We print NASA's handbooks and standards for the convenience of those that use them on a daily basis. We print all of these a full 8 ½ by 11 with large text so they are easy to read. Yes, color books are expensive to print so unless the information relies on the use of color for proper interpretation or understanding, we print most books in black and white to keep the cost down. All these documents are available for download for free from NASA, however printing them all over a network printer would take days.

Why buy a book you can download free? We print this so you don't have to.

All these books are available for free download from the government web site. Some are available only in electronic media. Some online docs are missing pages or barely legible.

We at 4th Watch Publishing are former government employees, so we know how government employees actually use the standards. When a new standard is released, an engineer prints it out, punches holes and puts it in a 3-ring binder. While this is not a big deal for a 5 or 10-page document, many NIST documents are over 100 pages and printing a large document is a time-consuming effort. So, an engineer that's paid $75 an hour is spending hours simply printing out the tools needed to do the job. That's time that could be better spent doing engineering. We publish these documents so engineers can focus on what they were hired to do – engineering. It's much more cost-effective to just order the latest version from Amazon.com

If there is a standard you would like published, let us know. Our web site is www.usgovpub.com

www.usgovpub.com

Copyright © 2019 4th Watch Publishing Co. All Rights Reserved

List of Other NASA Publications Available on Amazon.com:

NASA-STD-5001B	Structural Design and Test Factors of Safety for Spaceflight Hardware
NASA-STD-5006A	General Welding Requirements for Aerospace Materials
NASA-STD-5008B	Protective Coating of Carbon Steel, Stainless Steel, and Aluminum on Launch Structures, Facilities, and Ground Support Equipment
NASA-STD-5009A	Nondestructive Evaluation Requirements for Fracture-Critical Metallic Components
NASA-STD-5012B	Strength and Life Assessment Requirements for Liquid-Fueled Space Propulsion System Engines
NASA-STD-5019A	Fracture Control Requirements for Spaceflight Hardware
NASA-STD-5005D	Standard for The Design and Fabrication of Ground Support Equipment
NASA-HDBK-8739.21	Workmanship Manual for Electrostatic Discharge Control
NASA-HDBK 8739.23A	NASA Complex Electronics Handbook for Assurance Professionals (Color)
NASA-HDBK-8719.14	Handbook for Limiting Orbital Debris (Color)
NASA-HDBK-8709.22	Safety and Mission Assurance Acronyms, Abbreviations, and Definitions
NASA-HDBK-7009	NASA Handbook for Models and Simulations: An Implementation Guide For NASA-STD-7009 (Color)
NASA-HDBK-8739.19-2	Measuring and Test Equipment Specifications NASA Measurement Quality Assurance Handbook – Annex 2
NASA-HDBK-8739.19-3	Measurement Uncertainty Analysis Principles and Methods NASA Measurement Quality Assurance Handbook – Annex 3
NASA-HDBK-8739.19-4	Estimation and Evaluation of Measurement Decision Risk NASA Measurement Quality Assurance Handbook – Annex 4
NASA RCM	Reliability-Centered Maintenance Guide for Facilities and Collateral Equipment

www.usgovpub.com

Copyright © 2019 4th Watch Publishing Co. All Rights Reserved

METRIC/SI (ENGLISH)

NASA TECHNICAL STANDARD

National Aeronautics and Space Administration

NASA-STD-5001B
w/CHANGE 2:
ADMINISTRATIVE/
EDITORIAL CHANGES
2016-10-05

Approved: 2014-08-06
Superseding NASA-STD-5001A

STRUCTURAL DESIGN AND TEST FACTORS OF SAFETY FOR SPACEFLIGHT HARDWARE

APPROVED FOR PUBLIC RELEASE—DISTRIBUTION IS UNLIMITED

NASA-STD-5001B W/CHANGE 2

DOCUMENT HISTORY LOG

Status	Document Revision	Change Number	Approval Date	Description
Baseline			1996-6-21	Baseline Release
Interim Revision	A		2006-09-12	General Interim Revision
Revision	A		2008-08-05	General Revision Transition of Interim NASA Technical Standard NASA-STD-(I)-5001A to NASA Technical Standard NASA-STD-5001A.
Revision	B		2014-08-06	General Revision.
		1	2016-04-04	Revalidated w/Administrative/Editorial Changes—This NASA Technical Standard was reviewed and no technical changes resulted. Administrative changes to number requirements, add a Requirements Compliance Matrix as Appendix A, and conform to the current template were made, along with editorial corrections. The reference was moved to Appendix B.
		2	2016-10-05	Administrative/Editorial Changes—Added a "Caution Note" at the end of the guidance text in section 4.5, Buckling, based on issuance of the NASA Engineering and Safety Center Technical Bulletin No. 16-01, Buckling Knockdown Factors for Composite Cylinders.

APPROVED FOR PUBLIC RELEASE—DISTRIBUTION IS UNLIMITED

NASA-STD-5001B W/CHANGE 2

FOREWORD

This NASA Technical Standard is published by the National Aeronautics and Space Administration (NASA) to provide uniform engineering and technical requirements for processes, procedures, practices, and methods that have been endorsed as standard for NASA programs and projects, including requirements for selection, application, and design criteria of an item.

This NASA Technical Standard is approved for use by NASA Headquarters and NASA Centers and Facilities and may be cited in contract, program, and other Agency documents as a technical requirement. It may also apply to the Jet Propulsion Laboratory and other contractors only to the extent specified or referenced in applicable contracts.

Revision B of this NASA Technical Standard establishes additional design and test requirements for habitable modules and beryllium structures. It also clarifies the differences between prototype and protoflight test programs.

Requests for information should be submitted via "Feedback" at https://standards.nasa.gov. Requests for changes to this NASA Technical Standard should be submitted via MSFC Form 4657, Change Request for a NASA Engineering Standard.

Original Signed By:　　　　　　　　　　　　　　　*08/06/2014*

_____　　　　　_____

Ralph R. Roe, Jr.　　　　　　　　　　　　　　　　　Approval Date
NASA Chief Engineer

APPROVED FOR PUBLIC RELEASE—DISTRIBUTION IS UNLIMITED

NASA-STD-5001B W/CHANGE 2

TABLE OF CONTENTS

SECTION		PAGE
DOCUMENT HISTORY LOG		2
FOREWORD		3
TABLE OF CONTENTS		4
LIST OF APPENDICES		5
LIST OF TABLES		5
1.	**SCOPE**	**6**
1.1	Purpose	6
1.2	Applicability	6
1.3	Tailoring	7
1.4	Constraints and Preconditions	7
2.	**APPLICABLE DOCUMENTS**	**7**
2.1	General	7
2.2	Government Documents	8
2.3	Non-Government Documents	8
2.4	Order of Precedence	8
3.	**ACRONYMS AND DEFINITIONS**	**9**
3.1	Acronyms	9
3.2	Definitions	9
4.	**REQUIREMENTS**	**13**
4.1	Selection Criteria for Factors of Safety	13
4.1.1	Prototype versus Protoflight Approaches	13
4.1.2	Test Verification Criteria	14
4.1.3	Probabilistic Methods	16
4.2	Design and Test Factors of Safety	16
4.2.1	Metallic Structures	17
4.2.2	Threaded Fastening Systems	18
4.2.3	Composite/Bonded Structures	18
4.2.4	Glass/Ceramics	19
4.2.5	Pressurized Structures, Pressure Vessels, Pressurized Components, and Habitable Modules	21
4.2.6	Softgood Structures	22
4.3	Beryllium Structures	23
4.4	Fatigue and Creep	24
4.5	Buckling	24
4.6	Alternate Approaches	25

APPROVED FOR PUBLIC RELEASE—DISTRIBUTION IS UNLIMITED

LIST OF APPENDICES

APPENDIX		PAGE
1	Requirements Compliance Matrix	26
2	References	34

LIST OF TABLES

TABLE	TITLE	PAGE
1	Minimum Design and Test Factors for Metallic Structures	18
2	Minimum Design and Test Factors for Composite/Bonded Structures	19
3	Minimum Design and Test Factors for Glass/Ceramics in Robotic Applications	20
4	Minimum Design and Test Factors for Bonds in Glass/Ceramic Structures	21
5	Minimum Design and Test Factors for Habitable Modules, Doors, and Hatches	22
6	Minimum Design and Test Factors for Structural Softgoods	23
7	Minimum Design and Test Factors for Beryllium Structures	23

NASA-STD-5001B W/CHANGE 2

STRUCTURAL DESIGN AND TEST FACTORS OF SAFETY FOR SPACEFLIGHT HARDWARE

1. SCOPE

1.1 Purpose

The purpose of this NASA Technical Standard is to establish NASA structural design and test factors, as well as service life factors to be used for spaceflight hardware development and verification. The primary objective of this NASA Technical Standard is to define factors which ensure safe and reliable structural designs. The secondary objective is to reduce space project costs and schedules by enhancing the commonality of use of hardware designs among NASA flight projects, Centers, and their contractors. The criteria in this NASA Technical Standard are to be considered as minimum acceptable values unless adequate engineering risk assessment is provided that justifies the use of lower values.

1.2 Applicability

1.2.1 This NASA Technical Standard defines engineering practices for NASA programs and projects.

1.2.2 This NASA Technical Standard is approved for use by NASA Headquarters and NASA Centers and Facilities and may be cited in contract, program, and other Agency documents as a technical requirement. It may also apply to the Jet Propulsion Laboratory and other contractors only to the extent specified or referenced in applicable contracts.

1.2.3 Verifiable requirements are numbered and indicated by the word "shall"; this NASA Technical Standard contains 65 requirements. Explanatory or guidance text is indicated in italics beginning in section 4. To facilitate requirements selection and verification by NASA programs and projects, a Requirements Compliance Matrix is provided in Appendix A.

1.2.4 [FSR 1] NASA programs and projects that do not meet the provisions of this NASA Technical Standard shall be assessed by the NASA Program Manager for the associated risk to the success of the planned NASA mission and approved by the responsible Technical Authority.

1.2.5 [FSR 2] This NASA Technical Standard shall not supersede applicable laws and regulations unless a specific exemption has been obtained by the Office of the NASA Chief Engineer.

1.2.6 The criteria in this NASA Technical Standard are applicable to launch vehicle payloads and launch vehicle structures (including propellant tanks and solid rocket motor (SRM) cases). These criteria apply to flight hardware that is utilized for NASA missions. This NASA Technical Standard presents acceptable minimum factors of safety for use in analytical assessment and test verification of structural adequacy of the flight hardware. Designs are generally to be verified by both structural analyses and tests.

APPROVED FOR PUBLIC RELEASE—DISTRIBUTION IS UNLIMITED

1.2.7 Criteria are specified for design and test of flight articles when the actual flight hardware is tested (protoflight), and when qualification tests are conducted on a separate (prototype) article. In general, no distinction is made between "human-rated" and "robotic" missions. Structures of human-rated flight systems may be subjected to additional verification and/or safety requirements (e.g., fracture control) that are consistent with the established risk levels for mission success and flight crew safety.

1.2.8 Specifically excluded from this NASA Technical Standard are requirements for design loads determination and fracture control. Also excluded are the design and test factors for engines, rotating hardware, solid propellant, insulation, ground support equipment, and facilities. This NASA Technical Standard also does not cover specific configuration factors such as fitting factors, buckling knockdown factors, and load uncertainty factors.

1.3 Tailoring

[FSR 3] Tailoring of this NASA Technical Standard for application to a specific program or project shall be formally documented as part of program or project requirements and approved by the responsible Technical Authority in accordance with NPR 7120.5, NASA Space Flight Program and Project Management Requirements.

1.4 Constraints and Preconditions

The criteria of this NASA Technical Standard were developed in the context of structural and mechanical systems designs that are amenable to engineering analyses by current state-of-the-art methods and are in conformance with standard aerospace industry practices. More specifically, the designs are assumed to use materials having mechanical properties that are well characterized for the intended service environments and all design conditions. For reusable and multi-mission hardware, these criteria are applicable throughout the design service life and all of the missions.

1.4.1 [FSR 4] The service environments and limit loads shall be well defined.

1.4.2 [FSR 5] Aerospace standard manufacturing and process controls shall be used in hardware fabrication and handling.

1.4.3 [FSR 6] Deviations of the test article from the flight configuration shall be documented and approved by the responsible Technical Authority.

Test hardware should, as far as is practical, be representative of the flight configuration.

2. APPLICABLE DOCUMENTS

2.1 General

The documents listed in this section contain provisions that constitute requirements of this NASA Technical Standard as cited in the text.

2.1.1 [FSR 7] The latest issuances of cited documents shall apply unless specific versions are designated.

2.1.2 [FSR 8] Non-use of specifically designated versions shall be approved by the responsible Technical Authority.

The applicable documents are accessible at https://standards.nasa.gov, may be obtained directly from the Standards Developing Body or other document distributors, or information for obtaining the document is provided.

2.2   Government Documents

NPR 7120.5	NASA Space Flight Program and Project Management Requirements
NASA-STD-5018	Strength Design and Verification Criteria for Glass, Ceramics, and Windows in Human Space Flight Applications
NASA-STD-5020	Requirements for Threaded Fastening Systems in Spaceflight Hardware
NASA-STD-6016	Standard Materials and Processes Requirements for Spacecraft

2.3   Non-Government Documents

ANSI/AIAA S-080	Space Systems – Metallic Pressure Vessels, Pressurized Structures, and Pressure Components
ANSI/AIAA S-081	Space Systems – Composite Overwrapped Pressure Vessels (COPVs)

See Appendix B for references.

2.4   Order of Precedence

2.4.1 The requirements and standard practices established in this NASA Technical Standard do not supersede or waive existing requirements and standard practices found in other Agency documentation.

2.4.2 [FSR 9] Conflicts between this NASA Technical Standard and other requirements documents shall be resolved by the responsible Technical Authority.

3. ACRONYMS AND DEFINITIONS

3.1 Acronyms

AIAA	American Institute of Aeronautics and Astronautics
ANSI	American National Standards Institute
COPVs	composite overwrapped pressure vessels
FSR	factors of safety requirements
KDF	knock-down factor
kPa	kilopascal
MDP	maximum design pressure
MEOP	maximum expected operating pressure
N/A	not applicable
NASA	National Aeronautics and Space Administration
NESC	NASA Engineering and Safety Center
psi	pounds per square inch
psia	pounds per square inch absolute
SI	Systeme Internationale, or metric system of measurement
SRM	solid rocket motor

3.2 Definitions

Acceptance Test: A test performed to demonstrate that the hardware is acceptable for its intended use. It also serves as a quality control screen to detect manufacturing, material, or workmanship defects in the flight build and to demonstrate compliance with specified requirements. *Note: Acceptance tests are performed on previously qualified hardware to limit loading conditions.*

Creep: A time-dependent deformation under load and thermal environments that results in cumulative, permanent deformation.

Detrimental Yielding: Yielding that adversely affects the form, fit, and function, or integrity of the structure.

Discontinuity Area: A local region of a composite or non-metallic structure consisting of thickness changes, built-up plies, dropped plies, chopped fiber or reinforced regions around fittings, joints, or interfaces. In these regions, the stress state and load distribution may be difficult to characterize by analysis. *Note: Bonded joints are considered discontinuities.*

Factored Load or Stress: The limit load or stress multiplied by the appropriate design or test factor.

Factors of Safety (Safety Factors): Multiplying factors to be applied to limit loads or stresses for purposes of analytical assessment (design factors) or test verification (test factors) of design adequacy in strength or stability.

Failure: Rupture, collapse, excessive deformation, or any other phenomenon resulting in the inability of a structure to sustain specified loads, pressures, and environments or to function as designed.

Fatigue: The cumulative irreversible damage incurred in materials caused by cyclic application of stresses and environments, resulting in degradation of load-carrying capability.

Glass: Composed of any of a large class of materials with highly variable mechanical and optical properties that solidify from the molten state without crystallization and is typically made by fusing silicates with boric oxide, aluminum oxide, or phosphorus pentoxide; generally hard, brittle, and transparent or translucent; an amorphous (non-crystalline) material that is isotropic and elastic.

Habitable Module: A pressurized, life-supporting enclosure or module that is normally intended to support life without the need for spacesuits or a special breathing apparatus. The enclosure may be one that is continuously inhabited, or one that is used for crew transfer or crew-accessible stowage, as long as life support is a requirement for the design. Single mission or multi-mission designs are included.

Limit Load: The maximum anticipated load, or combination of loads that a structure may experience during its design service life under all expected conditions of operation.

Margin of Safety (MS): MS = [Allowable Load (Yield or Ultimate)/Limit Load*Factor of Safety (Yield or Ultimate)] - 1. *Note: Load may refer to force, stress, or strain.*

Maximum Design Pressure (MDP): The highest possible operating pressure considering maximum temperature, maximum relief pressure, maximum regulator pressure, and, where applicable, transient pressure excursions. MDP for human-rated hardware is a two-failure tolerant pressure; i.e., MDP will not be exceeded for any combination of two credible failures that will affect pressure. For all other hardware, MDP is equivalent to MEOP.

Maximum Expected Operating Pressure (MEOP): The maximum pressure which pressurized hardware is expected to experience during its service life, in association with its applicable operating environments. MEOP includes the effects of temperature, transient peaks, vehicle acceleration, and relief valve tolerance.

Pressure Vessel: A container designed primarily for storing pressurized gases or liquids and that:
(1) Contains stored energy of 19,309 Joules (14,240 ft-lb) or greater, based on adiabatic expansion of a perfect gas; or
(2) Will experience a maximum design pressure greater than 689.5 kiloPascal (kPa) absolute (100 pounds per square inch absolute (psia)); or
(3) Contains a pressurized fluid in excess of 103.4 kPa absolute (15 psia), which will create a safety hazard, if released.

Pressurized Component: A component in a pressurized system, other than a pressure vessel, pressurized structure, or special pressurized equipment that is designed largely by the internal pressure. Examples are lines, fittings, gauges, valves, bellows, and hoses.

Pressurized Structures: Structures designed to carry both internal pressure loads and vehicle structural loads. The main propellant tank of a launch vehicle is a typical example.

Proof Test: A test performed on flight hardware to screen for defects in workmanship and material quality, and to verify structural integrity. *Note: Proof tests are performed at a load or pressure in excess of limit load or MDP but below the yield strength of the hardware. Proof tests are performed on each flight unit for structures whose strength is workmanship or fabrication dependent. Proof tests are also used to screen for initial flaws in fracture critical items.*

Proof Test Factor: A multiplying factor to be applied to the limit load or MDP to define the proof test load or pressure.

Protoflight Hardware: Hardware that is qualified using a protoflight verification approach.

Protoflight Test: A test performed on flight or flight-like hardware (i.e., is built with same drawings, materials and processes as the flight unit) to demonstrate that the design meets structural integrity requirements. The test is performed at loads or pressure in excess of limit load or maximum design pressure but below the yield strength of the structure. When performed on flight structure, the test also verifies the workmanship and material quality of the flight build. *Note: Protoflight tests combine elements of prototype and acceptance test programs.*

Protoflight Test Factor: A multiplying factor to be applied to limit load or MDP to define the protoflight test load or pressure.

APPROVED FOR PUBLIC RELEASE—DISTRIBUTION IS UNLIMITED

Prototype Hardware: Hardware of a new design that is produced from the same drawings and using the same materials, tooling, manufacturing processes, inspection methods, and personnel competency levels as will be used for the flight hardware. *Note: Prototype hardware is dedicated test hardware that is not intended to be used as a flight unit.*

Prototype Test: A test conducted using prototype hardware to demonstrate that all structural integrity requirements have been met. *Note: Prototype testing is performed at load levels sufficient to demonstrate that the test article will not fail at ultimate design loads.*

Qualification Test: A test performed to qualify the hardware design for flight. *Note: Qualification tests are conducted on a flight-quality structure at load levels sufficient to demonstrate that all structural design requirements have been met. Both protoflight and prototype tests are considered qualification tests.*

Qualification Test Factor: A multiplying factor to be applied to the limit load or MDP to define the qualification test load or pressure.

Safety Critical: A classification for structures, components, procedures, etc., whose failure to perform as designed or produce the intended results would pose a threat of serious personal injury or loss of life.

Service Life: All significant loading cycles or events during the period beginning with manufacture of a component and ending with completion of its specified use. Testing, transportation, lift-off, ascent, on-orbit operations, descent, landing, and post-landing events are to be considered.

Service Life Factor (Life Factor): A multiplying factor to be applied to the maximum expected number of load cycles in the service life to determine the design adequacy in fatigue or fracture.

Special Pressurized Equipment: A piece of equipment that meets the pressure vessel definition, but which is not feasible or cost effective to comply with the requirements applicable to pressure vessels. Included are batteries, heat pipes, cryostats, and sealed containers.

Structural Softgoods: Straps, fabrics, inflatable structures, gossamer structures, and other similar structures that carry structural loads.

Threaded Fastening System (Fastening System): An assembled combination of a fastener, an internally threaded part, such as a nut or an insert, and also the region of all parts clamped between them, including washers, compressed by the fastener preload.

APPROVED FOR PUBLIC RELEASE—DISTRIBUTION IS UNLIMITED

NASA-STD-5001B W/CHANGE 2

<u>Ultimate Design Load</u>: The product of the ultimate factor of safety and the limit load.

<u>Ultimate Strength</u>: The maximum load or stress that a structure or material can withstand without incurring failure.

<u>Unfactored Load or Stress</u>: The limit load or stress before application of any design or test factors.

<u>Verification</u>: Any combination of test or analysis used to demonstrate that the hardware meets the defined requirements.

<u>Workmanship Verification</u>: Any test or inspection, including visual, dimensional, and non-destructive evaluation, performed to demonstrate the adequacy of the flight build. Tests performed to demonstrate workmanship may be static or dynamic.

<u>Yield Design Load</u>: The product of the yield factor of safety and the limit load.

<u>Yield Strength</u>: The maximum load or stress that a structure or material can withstand without incurring detrimental yielding.

4. REQUIREMENTS

4.1 Selection Criteria for Factors of Safety

The appropriate design and test factors for a given mechanical or structural flight hardware element depend on several parameters, such as the materials used, attachment methods (e.g., bonding), and the verification approach (prototype or protoflight). In addition to the minimum factors of safety specified in this NASA Technical Standard, some structural and mechanical members may be required to meet other more stringent and restrictive performance requirements, such as dimensional stability, pointing accuracy, stiffness/frequency constraints, or safety requirements (e.g., fracture control).

4.1.1 Prototype versus Protoflight Approaches

The standard accepted practice for verification of launch vehicles and human-rated spaceflight hardware is the prototype approach in which a separate, dedicated test structure, identical to the flight structure, is tested to ultimate loads to demonstrate that the design meets both yield and ultimate factor-of-safety requirements.

An acceptable alternative for verification of spacecraft and science payloads is the protoflight approach, wherein the flight structure is tested to levels above limit load but below yield strength to verify workmanship and demonstrate structural integrity of the flight hardware.

APPROVED FOR PUBLIC RELEASE—DISTRIBUTION IS UNLIMITED

The protoflight verification approach has the advantage that a dedicated test unit is not required, because qualification testing can be performed on the flight hardware. However, a protoflight verification approach does require that margin over flight limit loads be demonstrated by test; therefore, higher yield design factors of safety are required to prevent damage to the flight structure. Under a protoflight verification approach, yield and ultimate modes of failure or structural margins are not directly verified by test.

[FSR 10] A protoflight test shall be followed by inspection and functionality assessment.

Consideration should be given to development testing prior to committing to major test article configurations and especially prior to committing the flight article to protoflight test.

4.1.2 Test Verification Criteria

4.1.2.1 Test Methods

Strength verification tests fall into three basic categories: (1) tests to verify strength of the design (qualification); (2) tests to verify strength models; and (3) tests to screen for workmanship and material defects in the flight articles (acceptance or proof).

Strength verification tests are normally static load tests covering critical load conditions in the three orthogonal axes and, generally, can be classified as prototype or protoflight (see section 4.1.1).

In some cases, alternative test approaches (centrifuge, below resonance sine burst, saw tooth shock, etc.) are more effective in reproducing the critical load or environmental conditions and may be used in lieu of static testing if it can be demonstrated that the resulting loads in the test article are equivalent to or larger than the limit loads multiplied by the test factor.

    a. [FSR 11] The strength verification program shall be approved by the responsible Technical Authority.

    b. [FSR 12] The magnitude of the static test loads shall be equivalent to limit loads multiplied by the qualification, acceptance, or proof test factor.

    c. [FSR 13] Strength model verification, if required, shall be accomplished over the entire load range.

Strength model verification is normally performed as part of the strength verification testing.

Verification of the strength model over the entire load range is especially important if the response of the test article is expected to be nonlinear.

Strength model verification may not be required if the load path is easily determined and straightforward and the flight loads are well characterized.

    d. [FSR 14] The test article shall be instrumented to provide sufficient test data for correlation with the strength model.

    e. [FSR 15] Each habitable module, propellant tank, and SRM case shall be proof pressure tested.

    f. [FSR 16] Departures from test plans and procedures, including failures that occur during testing or are uncovered as part of post-test inspection, shall be documented by a non-conformance report per the approved quality assurance plan.

4.1.2.2 Test versus Design Factors of Safety

When using the prototype structural verification approach, the minimum ultimate design factors are the same as the required qualification test factors for both metallic and composite/bonded structures, except in the case of discontinuity areas of composite/bonded structures used in safety critical applications.

    a. [FSR 17] When using the prototype structural verification approach, metallic structures shall be verified to have no detrimental yielding at yield design load before testing to full qualification load levels.

    b. [FSR 18] When using the protoflight structural verification approach, design factors shall be specified to prevent detrimental yielding of the metallic structure or damage to the composite/bonded flight structure during test.

4.1.2.3 Test versus No-Test Options

Structural designs generally should be verified by analysis and by either prototype or protoflight strength testing. For metallic structures only, it may be permissible to verify structural integrity by analysis alone without strength testing.

    a. [FSR 19] Analysis shall be provided with an acceptable engineering rationale for the "no-test" option.

    b. [FSR 20] To use the "no-test" approach, project-specific criteria and rationale shall be developed for review and approval by the responsible Technical Authority.

Projects that propose to use the "no-test" approach generally have to use larger factors of safety than for tested hardware.

Increasing the design factors of safety does not by itself justify a "no-test" approach. Some examples of criteria on which to base such an approach are as follows:

- *The structural design is simple (e.g., statically determinate) with easily determined load paths; the design has been thoroughly analyzed for all critical load conditions; and there is a high confidence in the magnitude of all significant loading events.*

- *The structure is similar in overall configuration, design detail, build quality, and critical load conditions to a previous structure that was successfully test verified, with good correlation of test results to analytical predictions, and for which the same level of process control has been maintained.*

- *Development and/or component tests have been successfully completed on critical, difficult-to-analyze elements of the structure, and correlation of the analytical model to test results has been demonstrated.*

4.1.3 Probabilistic Methods

Current standard NASA structural verification criteria are deterministic, and experience has shown these deterministic criteria to be adequate. The probabilistic method uses knowledge (or assumptions) of the statistical variability of the design variables to select design criteria for achieving an overall success confidence level.

[FSR 21] Any proposed use of probabilistic criteria to supplement or as an alternative to deterministic factors of safety shall be approved by the responsible Technical Authority on an individual-case basis.

4.2 Design and Test Factors of Safety

a. [FSR 22] The design factors of safety and test factors of this NASA Technical Standard are the minimum required values for NASA spaceflight structures and shall be applied to the limit stress condition, including additive thermal or pressure stresses.

b. [FSR 23] If pressure or temperature has a relieving or stabilizing effect on the mode of failure, then for analysis or test of that failure mode, the unfactored stresses induced by temperature or the minimum expected pressure shall be used in conjunction with the factored stresses from all other loads.

Calculation of the portion of thermal stress or load which acts to relieve or stabilize stresses due to other applied loads is dependent on assumptions regarding boundary conditions and constraints between structural elements. In cases where a thermal stress or load acts to relieve or stabilize a failure mode, a conservative estimate (i.e., minimum value) of the portion of the thermal stress or load providing relief or stability should be used so as to not overestimate the beneficial effect of temperature.

c. [FSR 24] Material selection and derivation of material design allowables shall follow the requirements defined in NASA-STD-6016, Standard Materials and Processes Requirements for Spacecraft.

Material allowables are to be chosen to minimize the probability of structural failure due to material variability. Considerations when specifying material design allowables include accounting for degradation of material properties under service environments and performance of sufficient material tests to establish values with an appropriate statistical basis.

    d. [FSR 25] The factored stresses shall not exceed material allowable stresses (yield and ultimate) under the expected temperature, pressure, and other operating conditions.

    e. [FSR 26] The hardware shall be designed to preclude any detrimental yielding under limit loads and, where applicable, under protoflight or proof test loads.

    f. [FSR 27] Applications of design and test factors to the development and verification of a structure shall be accepted by the responsible Technical Authority only when all the constraints and preconditions specified in section 1.4 are met.

If the proof test is to be used for fracture control flaw screening, higher factors than those listed here may be required for proof testing.

Factors of safety on yield are not specified for composite/bonded structures, glass, and bonds for structural glass.

4.2.1 Metallic Structures

Spaceflight metallic structures can be developed using either the prototype or the protoflight approach.

    a. [FSR 28] The minimum design and test factors of safety for metallic structures shall be as specified in table 1.

    b. [FSR 29] Workmanship verification shall be performed for the first flight build and follow-on structures under a prototype test approach and for follow-on structures under a protoflight test approach.

    c. [FSR 30] The workmanship verification program shall be approved by the responsible Technical Authority.

Workmanship may be demonstrated by static or dynamic testing. Dynamic tests may be sinusoidal vibration, random vibration, or acoustic. The loads or responses generated in a workmanship test should be sufficient to verify the structural integrity of the flight hardware.

NASA-STD-5001B W/CHANGE 2

Table 1—Minimum Design and Test Factors for Metallic Structures

Verification Approach	Ultimate Design Factor	Yield Design Factor	Qualification Test Factor	Proof Test Factor
Prototype	1.4	1.0*	1.4	N/A or 1.05**
Protoflight	1.4	1.25	1.2	N/A or 1.05**

* Structure has to be assessed to prevent detrimental yielding during its design service life, acceptance, or proof testing.
** Propellant tanks and SRM cases only.

The strength qualification requirements under a protoflight verification approach may be satisfied with a dedicated flight-like qualification unit developed for strength testing but not intended to be used as a flight unit. For this type of strength qualification program, workmanship verification is required for the first flight build as specified in 4.2.1b.

4.2.2 Threaded Fastening Systems

a. [FSR 31] Threaded fastening systems shall be designed and analyzed per the requirements specified in NASA-STD-5020, Requirements for Threaded Fastening Systems in Spaceflight Hardware.

b. [FSR 32] The minimum design and test factors of safety for metallic fasteners and internally threaded parts shall be as specified in table 1.

The minimum design and test factors of safety for other components of a threaded fastening system not specified in 4.2.2b such as non-metallic fasteners, clamped parts or washers are covered by the appropriate section of this NASA Technical Standard.

4.2.3 Composite/Bonded Structures

[FSR 33] Composite/bonded structures, including bonded sandwich structures and bonded inserts but excluding glass, developed for NASA spaceflight missions, shall, as a minimum, use the design and test factors specified in table 2.

Metallic honeycomb (facesheets and core) is considered to be metallic structure.

For composite/bonded structures, each flight article has to be proof tested unless the requirements of section 4.6 are met.

APPROVED FOR PUBLIC RELEASE—DISTRIBUTION IS UNLIMITED

Table 2—Minimum Design and Test Factors for Composite/Bonded Structures

Verification Approach	Geometry of Structure	Ultimate Design Factor	Qualification Test Factor	Proof Test Factor
Prototype	Discontinuity Area	2.0*	1.4	1.05
	Uniform Material	1.4	1.4	1.05
Protoflight	Discontinuity Area	2.0*	1.2	1.2
	Uniform Material	1.5	1.2	1.2

* Factor applies to concentrated stresses. For nonsafety-critical applications, this factor may be reduced to 1.4 for prototype structures and 1.5 for protoflight structures.

4.2.4 Glass/Ceramics

Because of their brittle nature and susceptibility to moisture-assisted crack growth, glass and ceramic structures pose a special challenge for design and analysis. The strength of glass and ceramics found in literature can be misleading and is almost never applicable to NASA applications. Knowledge of fracture toughness, crack growth characteristics, and environments are required to fully understand the ability of the structure to withstand a given stress for the required time. Fracture toughness and crack growth rates are basic material properties of glass and ceramics that have to be used in conjunction with environmental exposure to determine the adequacy of the structure. The environmental exposure has four components:

- *Stress.*
- *Time.*
- *Moisture.*
- *Flaws/cracks.*

Glass and ceramic structures are always imperfect and contain flaws either inherent in the material or created from manufacturing or use. Therefore, for glass or ceramic structures such as windows that have to carry pressure loads for extended times, fracture mechanics principles are required to assess strength.

4.2.4.1 Glass/Ceramics in Human-Rated Spaceflight Applications

[FSR 34] The design and verification of glass and ceramic structures for human-rated spaceflight application shall follow the requirements defined in NASA-STD-5018, Strength Design and Verification Criteria for Glass, Ceramics, and Windows in Human Space Flight Applications.

APPROVED FOR PUBLIC RELEASE—DISTRIBUTION IS UNLIMITED

4.2.4.2 Glass/Ceramics in Robotic Spaceflight Applications

[FSR 35] A non-fracture based strength approach is possible for glass and ceramic structures in non-safety critical robotic spaceflight applications, such as mirrors and lenses used in science instruments, provided conservative strength allowables are used; however, approval by the responsible Technical Authority of the allowables and approach shall be obtained before implementing this option.

These options include, but are not limited to, the following:

- *Determining the allowable through a Weibull distribution.*
- *Proof testing the actual article and using this proof test value as the ultimate strength.*
- *Using a "low" initial ultimate strength allowable of 1000 pounds per square inch (psi) for glass.*
- *Using a test verified threshold stress for the particular type of glass/ceramic chosen.*

It should be noted that the traditional strength approach does not waive any fracture control requirements and that fracture-critical structures would still require a fracture mechanics assessment. Furthermore, the proof tests specified in this NASA Technical Standard are workmanship tests; fracture mechanics considerations may drive the project to a higher proof test factor.

a. [FSR 36] The minimum design and test factors for pressurized and nonpressurized glass/ceramics shall be as specified in table 3.

Table 3—Minimum Design and Test Factors for Glass/Ceramics in Robotic Applications

Verification Approach	Loading Condition	Ultimate Design Factor	Proof Test Factor
Test	Nonpressurized	3.0	1.2
	Pressurized	3.0	2.0
Analysis Only*	Nonpressurized	5.0	N/A

* Not applicable to ceramic structures.

b. [FSR 37] Structural integrity of all pressurized glass and ceramics shall be verified by both analysis and proof testing.

c. [FSR 38] Proof tests of glass and ceramics shall be configured to simulate flight-like boundary conditions and loading.

d. [FSR 39] For glass proof testing, the total time during unload shall be as short as possible and the whole test performed in an environment designed to minimize unrealistic flaw growth.

For pressurized glass/ceramics and nonpressurized ceramics, each flight article has to be proof tested unless the requirements of section 4.6 are met. For nonpressurized glass, proof testing is required, unless the requirements of section 4.6 are met, or if the unit can demonstrate a positive margin of safety using the analysis only "Ultimate Design Factor," as defined in table 3.

    e. [FSR 40] All glass and ceramic bonds shall be proof tested in the bonded flight assembly.

    f. [FSR 41] The design and test factors for structural bonds in glass or ceramics shall be as specified in table 4.

Table 4—Minimum Design and Test Factors for Bonds in Glass/Ceramic Structures

Application	Ultimate Design Factor	Proof Test Factor
Nonpressurized	1.5	1.2
Pressurized	3.0	2.0

4.2.5 Pressurized Structures, Pressure Vessels, Pressurized Components, and Habitable Modules

The design and analysis of pressurized structure is covered by this NASA Technical Standard. All relevant combinations of structural, thermal, and pressure loading are applicable.

4.2.5.1 Pressure Vessels and Pressurized Components

    a. [FSR 42] Metallic pressure vessels and pressurized components shall be designed, qualified, and accepted per the requirements of ANSI/AIAA S-080, Space Systems – Metallic Pressure Vessels, Pressurized Structures, and Pressurized Components.

    b. [FSR 43] Composite overwrapped pressure vessels (COPVs) shall be designed, qualified, and accepted per the requirements of ANSI/AIAA S-081, Space Systems – Composite Overwrapped Pressure Vessels (COPVs).

    c. [FSR 44] ANSI/AIAA S-080 and S-081 shall be tailored such that MDP is substituted for all references to maximum expected operating pressure (MEOP).

The term maximum design pressure (MDP) can be used for design and testing of pressure vessels and related pressure components. The basic difference between MDP and MEOP is the degree of consideration of potential credible failure within a pressure system and the resultant effects on pressure of the pressure vessel(s) during system operation. MDP is associated with human-rated systems and is based on the worst case combination of two credible system failures. For robotic hardware pressurization due to failure conditions are not included and the terms MDP and MEOP are equivalent.

APPROVED FOR PUBLIC RELEASE—DISTRIBUTION IS UNLIMITED

Habitable modules are not considered "pressure vessels."

4.2.5.2.1 Habitable Modules

a. [FSR 45] Habitable modules shall maintain dimensional stability required for functionality of structural and mechanical attachments, pressure connections, and openings for doors or hatches throughout their service life in the applicable environments.

b. [FSR 46] Habitable modules shall withstand applicable loads with the doors or hatches in the open and closed condition for the applicable ground and mission environments.

Habitable module structural integrity has to be maintained throughout all phases of service life for all hardware configurations, including conditions where the hatches and doors are opened or closed.

c. [FSR 47] The minimum design and test factors of safety for habitable modules, doors, and hatches shall be as specified in table 5.

Table 5—Minimum Design and Test Factors for Habitable Modules, Doors, and Hatches

Pressure Load Case	Yield Design Factor	Ultimate Design Factor	Proof Test Factor
Internal pressure only	1.65	2.0	1.5
Negative pressure differential*	N/A	1.5	N/A
Negative pressure differential if verified by analysis only	N/A	2.0	N/A

* Has to be capable of withstanding maximum external pressure multiplied by ultimate factor of safety (negative pressure differential) without collapse or rupture when internally pressurized to the minimum anticipated operating pressure.

For habitable modules, each flight article has to be proof tested unless the requirements of section 4.6 are met.

4.2.6 Softgood Structures

The design and test requirements for deployable decelerator systems such as parachutes, parafoils, airbags, inflatable heat shields and similar systems are not covered by this NASA Technical Standard.

4.2.6.1 [FSR 48] Static strength of all structural softgoods shall be test verified.

APPROVED FOR PUBLIC RELEASE—DISTRIBUTION IS UNLIMITED

4.2.6.2 [FSR 49] The minimum design and test factors of safety for structural softgoods shall be as specified in table 6.

Table 6—Minimum Design and Test Factors for Structural Softgoods

Hardware Criticality Classification	Ultimate Design Factor	Prototype Test Factor	Proof Test Factor
Loss of Life or Vehicle	4.0	4.0	1.2
All Others	2.0	2.0	1.2

Prototype testing is required for softgoods to demonstrate qualification. Each flight article has to be proof tested, unless the requirements of section 4.6 are met.

4.3 Beryllium Structures

A beryllium structure is any structure fabricated from materials that have a beryllium content of greater than 4 percent by weight as defined in NASA-STD-6016, Standard Materials and Processes Requirements for Spacecraft.

4.3.1 [FSR 50] The minimum design and test factors of safety for beryllium structures shall be as specified in table 7.

Table 7—Minimum Design and Test Factors for Beryllium Structures

Yield Design Factor	Ultimate Design Factor	Proof Test Factor
1.4	1.6	1.2

For beryllium structures, each flight article has to be proof tested unless the requirements of section 4.6 are met.

In addition to the design and test factors specified in table 7, the requirements given in sections 4.3.2 to 4.3.5 are also to be satisfied when using beryllium as a structural material.

4.3.2 [FSR 51] When using cross-rolled sheet, the design shall preclude out-of-plane loads and displacements during assembly, testing, or service life.

4.3.3 [FSR 52] Stress analysis shall properly account for the lack of ductility of the material by rigorous treatment of applied loads, boundary conditions, assembly stresses, stress concentrations, thermal cycling, possible material anisotropy, and worst-case tolerance conditions.

4.3.4 [FSR 53] All machined and/or mechanically disturbed surfaces shall be chemically milled to ensure removal of surface damage and residual stresses.

4.3.5 [FSR 54] All parts shall undergo penetrant inspection for surface cracks and crack-like flaws per NASA-STD-6016, Standard Materials and Processes Requirements for Spacecraft.

4.4 Fatigue and Creep

[FSR 55] For NASA spaceflight structures made of well-characterized materials and with sufficient load cycle data that accounts for all in-service environments, a minimum service life factor of 4.0 shall be applied to the service life for fatigue and creep life assessments.

For structures made of materials that are not well characterized or those that may have complex failure modes such as composite structures, an additional factor and testing may be required by the responsible Technical Authority.

4.5 Buckling

4.5.1 [FSR 56] All structural items subjected to significant in-plane stresses (compression and/or shear) under any combination of ground loads, flight loads, or thermal loads shall be analyzed for buckling failure.

4.5.2 [FSR 57] Design loads for buckling shall be ultimate loads.

4.5.3 [FSR 58] If a loading condition tends to alleviate buckling, then the unfactored load shall be used in combination with other factored loading conditions.

4.5.4 [FSR 59] Buckling evaluation shall address general instability, local or panel instability, crippling, and creep.

4.5.5 [FSR 60] Analyses of thin-walled shell structures subject to buckling load conditions during the service life shall account for the differences between idealized model geometry and the physical structure, including boundary conditions.

Discrepancies between analytically and empirically derived buckling load capability are due in part to the differences between idealized model geometry and the physical structure. "Knockdown factors" (correlation coefficients) are used to adjust predicted values to account for these differences. Typical knockdown factors for thin-walled circular cylinders are listed in NASA-SP-8007, Buckling of Thin-Walled Circular Cylinders. **Caution: *If NASA-SP-8007 is used to calculate a knock-down factor (KDF) for composite cylinders, the original assumptions and limitations should be understood and care should be taken. As noted in the NASA Engineering and Safety Center (NESC) Technical Bulletin No. 16-01, the universal KDF of 0.65 for composite shells may be unconservative for certain designs that are outside the scope of the publication. See NESC Technical Bulletin No. 16-01 for additional guidance.***

4.6 Alternate Approaches

4.6.1 [FSR 61] In the event a particular factor of safety requirement of this NASA Technical Standard cannot be met for a specific spaceflight structure or hardware component, an alternative or modified approach shall be proposed to verify the strength adequacy of the design.

4.6.2 [FSR 62] A written risk assessment that justifies the use of the alternate approach shall be prepared by the organization with primary responsibility for the development of the structure or component.

4.6.3 [FSR 63] The risk assessment shall be submitted to the responsible Technical Authority for approval prior to the implementation of the alternative approach.

4.6.4 [FSR 64] If the lower factors of safety are approved by the responsible Technical Authority, a waiver shall be written that documents the rationale for this one-time exception.

4.6.5 [FSR 65] Waivers shall not be used as precedents for future mission applications.

NASA-STD-5001B W/CHANGE 2

APPENDIX A

REQUIREMENTS COMPLIANCE MATRIX

A.1 Purpose

This Appendix provides a listing of requirements contained in this NASA Technical Standard for selection and verification of requirements by programs and projects. (*Note*: Enter "Yes" to describe the requirement's applicability to the program or project; or enter "No" if the intent is to tailor, and enter how tailoring is to be applied in the "Rationale" column.)

	NASA-STD-5001B w/CHANGE 1		
Section	Description	Requirement in this NASA Technical Standard	Applicable (Yes or No)	If No, Enter Rationale
1.2.4	Applicability	[FSR 1] NASA programs and projects that do not meet the provisions of this NASA Technical Standard shall be assessed by the NASA Program Manager for the associated risk to the success of the planned NASA mission and approved by the responsible Technical Authority.		
1.2.5	Applicability	[FSR 2] This NASA Technical Standard shall not supersede applicable laws and regulations unless a specific exemption has been obtained by the Office of the NASA Chief Engineer.		
1.3	Tailoring	[FSR 3] Tailoring of this NASA Technical Standard for application to a specific program or project shall be formally documented as part of program or project requirements and approved by the responsible Technical Authority in accordance with NPR 7120.5, NASA Space Flight Program and Project Management Requirements.		
1.4.1	Constraints and Preconditions	[FSR 4] The service environments and limit loads shall be well defined.		
1.4.2	Constraints and Preconditions	[FSR 5] Aerospace standard manufacturing and process controls shall be used in hardware fabrication and handling.		
1.4.3	Constraints and Preconditions	[FSR 6] Deviations of the test article from the flight configuration shall be documented and approved by the responsible Technical Authority.		
2.1.1	General	[FSR 7] The latest issuances of cited documents shall apply unless specific versions are designated.		
2.1.2	General	[FSR 8] Non-use of specifically designated versions shall be approved by the responsible Technical Authority.		
2.4.2	Order of Precedence	[FSR 9] Conflicts between this NASA Technical Standard and other requirements documents shall be resolved by the responsible Technical Authority.		
4.1.1	Prototype versus Protoflight	[FSR 10] A protoflight test shall be followed by inspection and functionality assessment.		

APPROVED FOR PUBLIC RELEASE—DISTRIBUTION IS UNLIMITED

NASA-STD-5001B W/CHANGE 2

Section	Description	NASA-STD-5001B w/CHANGE 1 Requirement in this NASA Technical Standard	Applicable (Yes or No)	If No, Enter Rationale
	Approaches			
4.1.2.1a	Test Methods	[FSR 11] The strength verification program shall be approved by the responsible Technical Authority.		
4.1.2.1b	Test Methods	[FSR 12] The magnitude of the static test loads shall be equivalent to limit loads multiplied by the qualification, acceptance, or proof test factor.		
4.1.2.1c	Test Methods	[FSR 13] Strength model verification, if required, shall be accomplished over the entire load range.		
4.1.2.1d	Test Methods	[FSR 14] The test article shall be instrumented to provide sufficient test data for correlation with the strength model.		
4.1.2.1e	Test Methods	[FSR 15] Each habitable module, propellant tank, and SRM case shall be proof pressure tested.		
4.1.2.1f	Test Methods	[FSR 16] Departures from test plans and procedures, including failures that occur during testing or are uncovered as part of post-test inspection, shall be documented by a non-conformance report per the approved quality assurance plan.		
4.1.2.2a	Test versus Design Factors of Safety	[FSR 17] When using the prototype structural verification approach, metallic structures shall be verified to have no detrimental yielding at yield design load before testing to full qualification load levels.		
4.1.2.2b	Test versus Design Factors of Safety	[FSR 18] When using the protoflight structural verification approach, design factors shall be specified to prevent detrimental yielding of the metallic structure or damage to the composite/bonded flight structure during test.		
4.1.2.3a	Test versus No-Test Options	[FSR 19] Analysis shall be provided with an acceptable engineering rationale for the "no-test" option.		
4.1.2.3b	Test versus No-Test Options	[FSR 20] To use the "no-test" approach, project-specific criteria and rationale shall be developed for review and approval by the responsible Technical Authority.		
4.1.3	Probabilistic Methods	[FSR 21] Any proposed use of probabilistic criteria to supplement or as an alternative to deterministic factors of safety shall be approved by the responsible Technical Authority on an individual-case basis.		
4.2a	Design and Test Factors of Safety	[FSR 22] The design factors of safety and test factors of this NASA Technical Standard are the minimum required values for NASA spaceflight structures and shall be applied to the limit stress condition, including additive thermal or pressure stresses.		
4.2b	Design and Test Factors of Safety	[FSR 23] If pressure or temperature has a relieving or stabilizing effect on the mode of failure, then for analysis or test of that failure mode, the unfactored stresses induced by temperature or the minimum expected pressure shall be used in conjunction with the factored stresses from all other loads.		
4.2c	Design and Test Factors of Safety	[FSR 24] Material selection and derivation of material design allowables shall follow the requirements defined in NASA-STD-6016Standard Materials and Processes Requirements for Spacecraft.		
4.2d	Design and Test Factors of Safety	[FSR 25] The factored stresses shall not exceed material allowable stresses (yield and ultimate) under the expected temperature, pressure, and other operating conditions.		
4.2e	Design and Test Factors of Safety	[FSR 26] The hardware shall be designed to preclude any detrimental yielding under limit loads and, where applicable, under protoflight or proof test loads.		
4.2f	Design and Test	[FSR 27] Applications of design and test factors to the development and verification of a structure shall be		

APPROVED FOR PUBLIC RELEASE—DISTRIBUTION IS UNLIMITED

NASA-STD-5001B W/CHANGE 2

Section	Description	Requirement in this NASA Technical Standard	Applicable (Yes or No)	If No, Enter Rationale
	Factors of Safety	accepted by the responsible Technical Authority only when all the constraints and preconditions specified in section 1.4 are met.		
4.2.1a	Metallic Structures	[FSR 28] The minimum design and test factors of safety for metallic structures shall be as specified in table 1. **Table 1—Minimum Design and Test Factors for Metallic Structures** 	Verification Approach	Ultimate Design Factor	Yield Design Factor	Qualification Test Factor	Proof Test Factor
Prototype	1.4	1.0*	1.4	N/A or 1.05**
Protoflight	1.4	1.25	1.2	N/A or 1.05**	 * Structure has to be assessed to prevent detrimental yielding during its design service life, acceptance, or proof testing. ** Propellant tanks and SRM cases only.		
4.2.1b	Metallic Structures	[FSR 29] Workmanship verification shall be performed for the first flight build and follow-on structures under a prototype test approach and for follow-on structures under a protoflight test approach.		
4.2.1c	Metallic Structures	[FSR 30] The workmanship verification program shall be approved by the responsible Technical Authority.		
4.2.2a	Threaded Fastening Systems	[FSR 31] Threaded fastening systems shall be designed and analyzed per the requirements specified in NASA-STD-5020, Requirements for Threaded Fastening Systems in Spaceflight Hardware.		
4.2.2b	Threaded Fastening Systems	[FSR 32] The minimum design and test factors of safety for metallic fasteners and internally threaded parts shall be as specified in table 1. **Table 1—Minimum Design and Test Factors for Metallic Structures** 	Verification Approach	Ultimate Design Factor	Yield Design Factor	Qualification Test Factor	Proof Test Factor
Prototype	1.4	1.0*	1.4	N/A or 1.05**			

NASA-STD-5001B W/CHANGE 2

Section	Description	Requirement in this NASA Technical Standard	Applicable (Yes or No)	If No, Enter Rationale
4.2.3	Composite/Bonded Structures	NASA-STD-5001B w/CHANGE 1 		1.4	1.25	1.2	
Protoflight				N/A or 1.05** | * Structure has to be assessed to prevent detrimental yielding during its design service life, acceptance, or proof testing. ** Propellant tanks and SRM cases only. [FSR 33] Composite/bonded structures, including bonded sandwich structures and bonded inserts but excluding glass, developed for NASA spaceflight missions, shall, as a minimum, use the design and test factors specified in table 2. **Table 2—Minimum Design and Test Factors for Composite/Bonded Structures** | Verification Approach	Geometry of Structure	Ultimate Design Factor	Qualification Test Factor	Proof Test Factor
Prototype	Discontinuity Area	2.0*	1.4	1.05
	Uniform Material	1.4	1.4	1.05
Protoflight	Discontinuity Area	2.0*	1.2	1.2
	Uniform Material	1.5	1.2	1.2	 * Factor applies to concentrated stresses. For nonsafety-critical applications, this factor may be reduced to 1.4 for prototype structures and 1.5 for protoflight structures.		
4.2.4.1	Glass/Ceramics in Human-Rated Spaceflight Applications	[FSR 34] The design and verification of glass and ceramic structures for human-rated spaceflight application shall follow the requirements defined in NASA-STD-5018, Strength Design and Verification Criteria for Glass, Ceramics, and Windows in Human Space Flight Applications.		
4.2.4.2	Glass/Ceramics in Robotic Spaceflight Applications	[FSR 35] A non-fracture based strength approach is possible for glass and ceramic structures in non-safety critical robotic spaceflight applications, such as mirrors and lenses used in science instruments, provided conservative strength allowables are used; however, approval by the responsible Technical Authority of the allowables and approach shall be obtained before implementing this option.		

NASA-STD-5001B W/CHANGE 2

Section	Description	Requirement in this NASA Technical Standard	Applicable (Yes or No)	If No, Enter Rationale
4.2.4.2a	Glass/Ceramics in Robotic Spaceflight Applications	[FSR 36] The minimum design and test factors for pressurized and nonpressurized glass/ceramics shall be as specified in table 3. **Table 3—Minimum Design and Test Factors for Glass/Ceramics in Robotic Applications** | Verification Approach	Loading Condition	Ultimate Design Factor	Proof Test Factor
---	---	---	---
Test	Nonpressurized	3.0	1.2
Test	Pressurized	3.0	2.0
Analysis Only*	Nonpressurized	5.0	N/A | * Not applicable to ceramic structures.		
4.2.4.2b	Glass/Ceramics in Robotic Spaceflight Applications	[FSR 37] Structural integrity of all pressurized glass and ceramics shall be verified by both analysis and proof testing.		
4.2.4.2c	Glass/Ceramics in Robotic Spaceflight Applications	[FSR 38] Proof tests of glass and ceramics shall be configured to simulate flight-like boundary conditions and loading.		
4.2.4.2d	Glass/Ceramics in Robotic Spaceflight Applications	[FSR 39] For glass proof testing, the total time during unload shall be as short as possible and the whole test performed in an environment designed to minimize unrealistic flaw growth.		
4.2.4.2e	Glass/Ceramics in Robotic Spaceflight Applications	[FSR 40] All glass and ceramic bonds shall be proof tested in the bonded flight assembly.		
4.2.4.2f	Glass/Ceramics in Robotic Spaceflight	[FSR 41] The design and test factors for structural bonds in glass or ceramics shall be as specified in table 4. **Table 4—Minimum Design and Test Factors for Bonds in Glass/Ceramic Structures**		

NASA-STD-5001B W/CHANGE 2

Section	Description	Requirement in this NASA Technical Standard (NASA-STD-5001B w/CHANGE 1)			Applicable (Yes or No)	If No, Enter Rationale
	Applications	Application	Ultimate Design Factor	Proof Test Factor		
		Nonpressurized	1.5	1.2		
		Pressurized	3.0	2.0		
4.2.5.1a	Pressure Vessels and Pressurized Components	[FSR 42] Metallic pressure vessels and pressurized components shall be designed, qualified, and accepted per the requirements of ANSI/AIAA S-080, Space Systems – Metallic Pressure Vessels, Pressurized Structures, and Pressurized Components.				
4.2.5.1b	Pressure Vessels and Pressurized Components	[FSR 43] Composite overwrapped pressure vessels (COPVs) shall be designed, qualified, and accepted per the requirements of ANSI/AIAA S-081, Space Systems – Composite Overwrapped Pressure Vessels (COPVs).				
4.2.5.1c	Pressure Vessels and Pressurized Components	[FSR 44] ANSI/AIAA S-080 and S-081 shall be tailored such that MDP is substituted for all references to maximum expected operating pressure (MEOP).				
4.2.5.2.1a	Habitable Modules	[FSR 45] Habitable modules shall maintain dimensional stability required for functionality of structural and mechanical attachments, pressure connections, and openings for doors or hatches throughout their service life in the applicable environments.				
4.2.5.2.1b	Habitable Modules	[FSR 46] Habitable modules shall withstand applicable loads with the doors or hatches in the open and closed condition for the applicable ground and mission environments.				
4.2.5.2.1c	Habitable Modules	[FSR 47] The minimum design and test factors of safety for habitable modules, doors, and hatches shall be as specified in table 5.				

Table 5—Minimum Design and Test Factors for Habitable Modules, Doors, and Hatches

Pressure Load Case	Yield Design Factor	Ultimate Design Factor	Proof Test Factor
Internal pressure only	1.65	2.0	1.5
Negative pressure differential*	N/A	1.5	N/A

APPROVED FOR PUBLIC RELEASE—DISTRIBUTION IS UNLIMITED

NASA-STD-5001B W/CHANGE 2

Section	Description	Requirement in this NASA Technical Standard	Applicable (Yes or No)	If No, Enter Rationale
		Negative pressure differential if verified by analysis only	N/A	2.0	N/A		
		* Has to be capable of withstanding maximum external pressure multiplied by ultimate factor of safety (negative pressure differential) without collapse or rupture when internally pressurized to the minimum anticipated operating pressure.		
4.2.6.1	Softgood Structures	[FSR 48] Static strength of all structural softgoods shall be test verified.		
4.2.6.2	Softgood Structures	[FSR 49] The minimum design and test factors of safety for structural softgoods shall be as specified in table 6. **Table 6—Minimum Design and Test Factors for Structural Softgoods** 	Hardware Criticality Classification	Ultimate Design Factor	Prototype Test Factor	Proof Test Factor
---	---	---	---
Loss of Life or Vehicle	4.0	4.0	1.2
All Others	2.0	2.0	1.2			
4.3.1	Beryllium Structures	[FSR 50] The minimum design and test factors of safety for beryllium structures shall be as specified in table 7. **Table 7—Minimum Design and Test Factors for Beryllium Structures** | Yield Design Factor	Ultimate Design Factor	Proof Test Factor
---	---	---
1.4	1.6	1.2			
4.3.2	Beryllium Structures	[FSR 51] When using cross-rolled sheet, the design shall preclude out-of-plane loads and displacements during assembly, testing, or service life.		
4.3.3	Beryllium Structures	[FSR 52] Stress analysis shall properly account for the lack of ductility of the material by rigorous treatment of applied loads, boundary conditions, assembly stresses, stress concentrations, thermal cycling, possible material anisotropy, and worst-case tolerance conditions.		
4.3.4	Beryllium Structures	[FSR 53] All machined and/or mechanically disturbed surfaces shall be chemically milled to ensure removal of surface damage and residual stresses.		

APPROVED FOR PUBLIC RELEASE—DISTRIBUTION IS UNLIMITED

NASA-STD-5001B W/CHANGE 2

	NASA-STD-5001B w/CHANGE 1		
Section	Description	Requirement in this NASA Technical Standard	Applicable (Yes or No)	If No, Enter Rationale
4.3.5	Beryllium Structures	[FSR 54] All parts shall undergo penetrant inspection for surface cracks and crack-like flaws per NASA-STD-6016, Standard Materials and Processes Requirements for Spacecraft.		
4.4	Fatigue and Creep	[FSR 55] For NASA spaceflight structures made of well-characterized materials and with sufficient load cycle data that accounts for all in-service environments, a minimum service life factor of 4.0 shall be applied to the service life for fatigue and creep life assessments.		
4.5.1	Buckling	[FSR 56] All structural items subjected to significant in-plane stresses (compression and/or shear) under any combination of ground loads, flight loads, or thermal loads shall be analyzed for buckling failure.		
4.5.2	Buckling	[FSR 57] Design loads for buckling shall be ultimate loads.		
4.5.3	Buckling	[FSR 58] If a loading condition tends to alleviate buckling, then the unfactored load shall be used in combination with other factored loading conditions.		
4.5.4	Buckling	[FSR 59] Buckling evaluation shall address general instability, local or panel instability, crippling, and creep.		
4.5.5	Buckling	[FSR 60] Analyses of thin-walled shell structures subject to buckling load conditions during the service life shall account for the differences between idealized model geometry and the physical structure, including boundary conditions.		
4.6.1	Alternate Approaches	[FSR 61] In the event a particular factor of safety requirement of this NASA Technical Standard cannot be met for a specific spaceflight structure or hardware component, an alternative or modified approach shall be proposed to verify the strength adequacy of the design.		
4.6.2	Alternate Approaches	[FSR 62] A written risk assessment that justifies the use of the alternate approach shall be prepared by the organization with primary responsibility for the development of the structure or component.		
4.6.3	Alternate Approaches	[FSR 63] The risk assessment shall be submitted to the responsible Technical Authority for approval prior to the implementation of the alternative approach.		
4.6.4	Alternate Approaches	[FSR 64] If the lower factors of safety are approved by the responsible Technical Authority, a waiver shall be written that documents the rationale for this one-time exception.		
4.6.5	Alternate Approaches	[FSR 65] Waivers shall not be used as precedents for future mission applications.		

APPROVED FOR PUBLIC RELEASE—DISTRIBUTION IS UNLIMITED

APPENDIX B

REFERENCES

B.1 Purpose

This Appendix provides reference information.

B.2 Reference Document

NASA Engineering and Safety Center (NESP) Technical Bulletin No. 16-01, Buckling Knock-Down Factors for Composite Cylinders

NASA-SP-8007, Buckling of Thin-Walled Circular Cylinders

www.ingramcontent.com/pod-product-compliance
Lightning Source LLC
Chambersburg PA
CBHW081648220526
45468CB00009B/2588